Let's Subtract
Bills

Kelly Doudna

Consulting Editor Monica Marx, M.A./Reading Specialist

Published by SandCastle™, an imprint of ABDO Publishing Company, 4940 Viking Drive, Edina, Minnesota 55435.

Copyright © 2003 by Abdo Consulting Group, Inc. International copyrights reserved in all countries. No part of this book may be reproduced in any form without written permission from the publisher. SandCastle™ is a trademark and logo of ABDO Publishing Company. Printed in the United States.

Credits
Edited by: Pam Price
Curriculum Coordinator: Nancy Tuminelly
Cover and Interior Design and Production: Mighty Media
Photo Credits: Comstock, Hemera, PhotoDisc, Stockbyte

Library of Congress Cataloging-in-Publication Data

Doudna, Kelly, 1963-
    Let's subtract bills / Kelly Doudna.
      p. cm. -- (Dollars & cents)
    Includes index.
    Summary: Shows how to use subtraction to find out how many dollars one has left after paying for various items and looks at different denominations of dollar bills, from one to one hundred.
    ISBN 1-57765-899-X
    1. Money--Juvenile literature. 2. Subtraction--Juvenile literature. [1. Money. 2. Subtraction.] I. Title. II. Series.

HG221.5 .D655 2002
640'.42--dc21
                                                                                                            2002071185

SandCastle™ books are created by a professional team of educators, reading specialists, and content developers around five essential components that include phonemic awareness, phonics, vocabulary, text comprehension, and fluency. All books are written, reviewed, and leveled for guided reading, early intervention reading, and Accelerated Reader® programs and designed for use in shared, guided, and independent reading and writing activities to support a balanced approach to literacy instruction.

Let Us Know

After reading the book, SandCastle would like you to tell us your stories about reading. What is your favorite page? Was there something hard that you needed help with? Share the ups and downs of learning to read. We want to hear from you! To get posted on the ABDO Publishing Company Web site, send us email at:

sandcastle@abdopub.com

SandCastle Level: Transitional

Bills are money.

one dollar
$1.00

five dollars
$5.00

ten dollars
$10.00

twenty dollars
$20.00

fifty dollars
$50.00

one hundred dollars
$100.00

We use bills to pay for things.

Let's see what we can buy.

Ken has 6 one-dollar bills.

The beach ball costs $2.00.
$2.00 = 2 one-dollar bills

How many one-dollar bills will he have left?

Let's subtract.
6 - 2 = 4

The robot costs $6.00.
$6.00 = 6 one-dollar bills

Brenda has 2 one-dollar bills.

How many more one-dollar bills does she need?

Let's subtract.
6 - 2 = 4

Liz has 5 five-dollar bills.

The teddy bear costs $10.00.
$10.00 = 2 five-dollar bills

How many five-dollar bills will she have left?

Let's subtract.
5 - 2 = 3

The wagon costs $25.00.
$25.00 = 5 five-dollar bills

Lin has 2 five-dollar bills.

How many more five-dollar bills does she need?

Let's subtract.
5 - 2 = 3

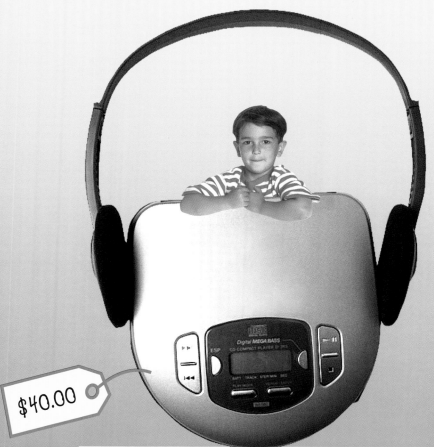

$40.00

The CD player costs $40.00.
$40.00 = 4 ten-dollar bills

Dave has 2 ten-dollar bills.

How many more ten-dollar bills does he need?

Let's subtract.
4 - 2 = 2

Al has 4 twenty-dollar bills.

The backpack costs $40.00.
$40.00 = 2 twenty-dollar bills

How many twenty-dollar bills will he have left?

Let's subtract.
4 - 2 = 2

Jen has 3 hundred-dollar bills.

The camera costs $200.00.
$200.00 = 2 hundred-dollar bills

How many hundred-dollar bills will she have left?

Let's subtract.
3 - 2 = 1

What are these bills called?

How much are they worth?

one dollar = $1.00
five dollars = $5.00
ten dollars = $10.00
twenty dollars = $20.00
fifty dollars = $50.00
one hundred dollars = $100.00

Index

backpack, p. 17
beach ball, p. 7
bills, pp. 3, 5, 21
camera, p. 19
CD player, p. 15
five-dollar bills,
   pp. 11, 13
hundred-dollar
   bills, p. 19
money, p. 3

one-dollar bills,
   pp. 7, 9
robot, p. 9
teddy bear, p. 11
ten-dollar bills,
   p. 15
twenty-dollar bills,
   p. 17
wagon, p. 13

Glossary

backpack   a bag you wear on your back

camera   a device that records images on film or electronically

CD player   a machine for playing compact discs

robot   a machine that performs jobs usually done by a human

teddy bear   a stuffed toy bear

About SandCastle™

A professional team of educators, reading specialists, and content developers created the SandCastle™ series to support young readers as they develop reading skills and strategies and increase their general knowledge. The SandCastle™ series has four levels that correspond to early literacy development in young children. The levels are provided to help teachers and parents select the appropriate books for young readers.

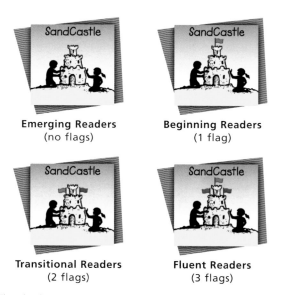

Emerging Readers
(no flags)

Beginning Readers
(1 flag)

Transitional Readers
(2 flags)

Fluent Readers
(3 flags)

These levels are meant only as a guide. All levels are subject to change.

To see a complete list of SandCastle™ books and other nonfiction titles from ABDO Publishing Company, visit www.abdopub.com or contact us at:
4940 Viking Drive, Edina, Minnesota 55435 • 1-800-800-1312 • fax: 1-952-831-1632